A Survival Guide:

A PM & Super Love Story

By John Abraham

Construction Superintendent at T&W Corporation

Copyright © 2024 John Abraham

All rights reserved. No part of this publication may be reproduced, distributed, or transmitted in any form or by any means, including photocopying, recording, or other electronic or mechanical methods, without the prior written permission of the publisher, except in the case of brief quotations embodied in critical reviews and certain other noncommercial uses permitted by copyright law. For permission requests, please contact the publisher at the email address below.

jabraham@twcorp.net

Table of Contents

1. The Importance of Strong Collaboration (Page 7)

2. Understanding Roles and Responsibilities (Page 13)

3. Common Challenges in PM-Super Relationships (Page 19)

4. Effective Communication Strategies (Page 25)

5. Aligning Goals and Expectations (Page 31)

6. Building Trust and Mutual Respect (Page 37)

7. Managing Budget, Time, and Quality Conflicts (Page 41)

8. Team Collaboration: Creating a Unified Leadership (Page 47)

9. Technology in Project Management (Page 53)

10. Problem Solving and Decision-Making (Page 59)

11. Conflict Resolution and Mediation (Page 65)

12. Leadership Development for PMs and Supers (Page 71)

13. Building Long-Term Relationships (Page 77)

14. Towards Better Project Outcomes (Page 83)

Dedication

"Employers utilize the skills of their employees; leaders develop and uplift those who follow them."

I dedicate this book to those who have shaped my journey as a superintendent. To Steve Shehorn, T&W CEO; Andrew Huey, T&W President; and Mike Updike, T&W DFO – thank you for fostering an environment where growth and excellence thrive.

My gratitude goes especially to my senior superintendent mentors, Steve Davis and Terry Kratz, who, like iron sharpening iron, have guided me with their wisdom and unwavering support.

And of course, it would be remiss not to acknowledge the project managers who've enriched my career in profound ways. I am truly blessed to have worked alongside Mike Priddy and Bryan Crostreet, two senior project managers who influenced me both professionally and spiritually.

Introduction: Why PMs and Supers Are the Construction World's Ultimate Power Couple

In the chaotic yet rewarding world of construction, the Project Manager (PM) and the Construction Superintendent (Super) have a relationship that could make or break a project. Think of them as the peanut butter and jelly of the construction sandwich—distinct flavors, but together they're unbeatable. While the PM plots the grand master plan, the Super is down in the trenches making it all happen. And just like any great couple, their success depends on how well they communicate, compromise, and maybe even occasionally vent about the chaos of a construction site.

Let's take a closer look at the dynamic duo and why their relationship is more important than your coffee breaks (well, almost).

Overview of Roles

- **Project Managers:** PMs are like the visionaries of the construction world, juggling plans, budgets, and schedules like circus performers with too many flaming swords. They coordinate resources, smooth-talk clients, and ensure that the project stays financially sound. They're the ones who dream big and expect those dreams to magically manifest within budget and on time—while also making sure every subcontractor, architect, and client is happy.

- **Construction Superintendents:** Meanwhile, Supers are the ones who turn those PM dreams into reality. They're the bosses of the job site, herding workers, managing chaos, solving real-time problems, and making sure the work gets done safely (without anyone losing a finger). While the PM is busy balancing spreadsheets, the Super is busy balancing beams—and probably trying to figure out how to explain why it rained on a pour day. They're the heart of the operation, making sure everything stays on track, even when the track is a muddy construction site.

Why Their Relationship is Critical

Much like a buddy-cop movie, the PM and Super need each other to win the day, catch the bad guy (budget overruns, anyone?), and live to see the project's completion. Here's why a strong working relationship between these two is essential:

- **Efficient Communication:** Communication between PMs and Supers is like Wi-Fi—when it's working, everything's great. When it's not, disaster. Without clear communication, the project can quickly descend into finger-pointing chaos, missed deadlines, and an unplanned budget funeral. Keeping that line of communication open means everyone knows what's happening—whether it's a schedule change, a scope shift, or that the port-a-potty tipped over in the wind again.

- **Aligning Plans with Reality:** PMs may have grand designs in mind, but it's the Super who has to turn those paper dreams into brick-and-mortar realities. If there's a disconnect between what's on paper and what can actually be done, well... you're in for a rough time. The Super's job is to keep the PM grounded in reality, while the PM keeps the Super thinking big picture. Together, they're like the perfect blend of pragmatism and idealism—or at least, they try to be.

- **Problem-Solving and Risk Mitigation:** Construction sites are like a never-ending episode of *Survivor*—problems and surprises pop up constantly. A great PM-Super combo can roll with the punches, strategize solutions, and keep the project moving, all while dodging potential disasters like budget explosions or weather-induced delays.

- **Meeting Client Expectations:** Clients expect miracles, and let's face it, PMs and Supers are basically construction magicians. The PM handles the client charm offensive, while the Super ensures the work meets expectations on the ground. Together, they keep clients happy and avoid awkward conversations about why the project's two weeks behind schedule (hint: it was the rain, again).

Common Tensions Between the Two Roles

Of course, it's not all sunshine and steel beams in the PM-Super world. They often butt heads, like siblings squabbling over who's hogging the remote, only in this case, it's about deadlines, budgets, and the occasional missing contractor.

- **Different Focuses:** The PM is obsessed with the project's bigger picture—deadlines, budgets, and client satisfaction. The Super, meanwhile, is laser-focused on getting the job done right on-site, here and now. This can cause friction, with the PM thinking the Super is too bogged down in details, and the Super thinking the PM is living in fantasyland.

- **Perception of Responsibilities:** PMs might think Supers are a bit too resistant to change, while Supers think PMs have never held a hammer in their lives. Each side feels misunderstood—PMs wish Supers would be more flexible, and Supers wish PMs would get their hands dirty (at least metaphorically).

- **Pressure from Above and Below:** PMs are getting heat from upper management or clients to stick to timelines and budgets, while Supers are managing a construction site full of real-world challenges like weather, material delays, and

Murphy's Law. It's like they're both holding a different end of the stress stick, pulling in opposite directions.

How to Improve Collaboration

Like any good relationship, the key to success is trust, communication, and maybe a little patience. PMs and Supers need to respect each other's roles and recognize that both are essential for a successful project. By working together, they can build a project (and a relationship) that stands the test of time—much like the buildings they construct.

In the following sections, we'll get into the nitty-gritty of how these two powerhouses can work together more effectively, avoid unnecessary drama, and keep things running smoothly—so maybe they can grab that coffee break after all.

Understanding Roles and Responsibilities: A PM and Super Story

In the world of construction, keeping roles and responsibilities clear between the Project Manager and the Construction Superintendent is like maintaining peace between two stubborn siblings. If both understand and respect each other's turf, things run smoothly. But when they don't—well, let's just say you're headed for a construction showdown that no one wants to see. So, buckle up as we break down the jobs of the PM and Super—and how things can go hilariously wrong when they misunderstand each other.

The Project Manager's Responsibilities:

The PM is basically the conductor of this chaotic orchestra. They're the mastermind with spreadsheets, highlighters, and a mysterious ability to see into the future (at least when it comes to budgets). Here's what they do:

1. **Budgeting:**
    - PMs are the ones who figure out how to get the job done without blowing the budget. They manage the dollars, negotiate with suppliers, and make sure that by the end, there's at least a few bucks left over for everyone to celebrate with coffee. The Super,

however, sometimes thinks they're pinching pennies where it counts.

2. **Client Communication:**
    - When the client wants to know how their precious building is coming along, the PM is there to smile, nod, and say, "Everything is going great!" even when the concrete just cracked, and it's raining sideways. The Super? Probably doing damage control on the ground while the PM's sugarcoating.

3. **Scheduling:**
    - PMs create grand schedules with color-coded charts and think, "Nothing can go wrong!" Meanwhile, the Super is battling real-life dragons like weather delays, supply chain issues, and the occasional missing worker. Deadlines? Ha!

4. **Contracts and Legal Agreements:**
    - Contracts? That's the PM's playground. They love nothing more than signing dotted lines and making sure everyone's legally bound to do their job. The Super is more focused on binding the building together with some actual steel, but hey, to each their own.

The Superintendent's Responsibilities:

The Super is the boots-on-the-ground commander. If the PM is the conductor, the Super's out there playing the instruments, building the stage, and making sure no one trips over the drum kit. Here's their job:

1. **On-Site Management:**
    - Supers make sure things actually get built—imagine that! They direct the workforce, solve problems in real-time, and occasionally wish they had a mute button for the PM's budget-cutting suggestions. Their goal is simple: build the thing without it collapsing.

2. **Quality Control:**
    - The Super's role is to keep things from falling apart—literally. They're the watchdogs for quality, ensuring that what's on the blueprint looks just as good (or at least close enough) in real life. If the PM tries to cheap out, the Super's there to push back with a knowing look and a very firm "no."

3. **Safety Management:**
    - Supers are the kings and queens of safety. They keep everyone in hard hats, enforce the rules, and remind

workers that, no, they can't ride the backhoe for fun. The PM appreciates this, but also wonders why everything seems to take twice as long when people aren't allowed to sprint across beams.

4. **Labor Coordination:**
    - Managing workers is like herding cats, but somehow the Super pulls it off. They make sure the right people are in the right place, doing the right thing—or at least close enough that nobody's called OSHA. PMs just want the job done, but Supers? They want it done *right*.

Where It All Goes Wrong: Misunderstandings

Now, here's where things get spicy.

1. **Budget vs. Site Realities:**
    - The PM thinks: "We can definitely save money by using cheaper materials." The Super, covered in dust and already thinking about future repairs, shoots back: "Yeah, sure, if you want a roof that leaks next month." And thus begins the great budget battle of every project ever.

2. **Scheduling Tensions:**
    - PM: "We're ahead of schedule!" Super: "We're what?" Turns out, the PM didn't notice that rain, sick workers, and a supplier mix-up mean they're actually behind by two weeks. Cue the frantic phone calls.

3. **Client Requests vs. Reality:**
    - PM: "The client wants an extra floor." Super: "The foundation can barely hold the building as it is." The PM thinks it's no big deal, but the Super knows that physics isn't just a suggestion.

4. **Authority Confusion:**
    - PM makes a call without consulting the Super. Super makes an on-site decision without telling the PM. Suddenly, no one's sure who's in charge, and the crew is taking orders from both. Chaos ensues.

Conclusion:

If you want to avoid turning your construction project into a sitcom, here's the trick: PMs and Supers need to talk, respect each other's roles, and maybe share a few cups of coffee (or strong tea). They both bring something essential to the table—whether it's keeping the budget on track or making sure the building doesn't collapse. When they work together, it's magic. When they don't? Well, at least you'll have some funny stories to tell at the next company party!

Common Challenges in PM-Super Relationships

Ah, the eternal struggle: Project Manager vs. Superintendent. It's like the construction version of Batman and Robin—except sometimes Batman doesn't appreciate Robin's site-based brilliance, and Robin thinks Batman's spreadsheets are a bit... much. Let's dive into some of the common challenges these dynamic duos face and how they can find harmony on the construction battlefield.

1. Communication Breakdowns

Communication is everything—except when it's nothing. PMs and Supers can have totally different conversations while staring right at each other. The PM's talking about budgets and deadlines, while the Super is wondering why the crane isn't moving.

- **Lack of Clarity:** PMs love to talk big picture: budgets, timelines, client expectations. Meanwhile, Supers need to know whether the materials are showing up today or if they'll be staring at a pile of dirt for the next six hours. It's like one person's reading a novel, and the other's stuck with an IKEA instruction manual.

- **Over-communication vs. Under-communication:** Some PMs will give you hourly updates that could rival your grandma's Facebook posts. On the flip side, some Supers are like covert agents—getting radio silence until something's on fire (literally or figuratively). Finding the sweet spot between "constant pinging" and "total blackout" is key to keeping everyone happy (and sane).

2. Differing Priorities

Ah yes, the classic showdown: What's more important, money or time? Budget or quality? It's like asking whether you prefer coffee or sleep—you need both to survive, but there's never enough of either.

- **Budget vs. On-Time Delivery:** The PM wants to save every penny—cut the labor, find cheaper materials, maybe even swap that crane for a pulley system? Meanwhile, the Super's trying to figure out how to get the project done with whatever magic beans are left. Spoiler alert: It doesn't always end well.

- **Quality vs. Speed:** Supers care about quality because, well, they don't want to be rebuilding the same wall three times. PMs, on the other hand, have deadlines so tight they've basically set up camp in the future. Finding the balance between "do it fast" and "do it right" can feel like a game of Jenga—pull the wrong block, and it all comes crashing down.

3. Trust and Respect Issues

The PM and Super relationship is a delicate dance. Too bad it sometimes feels like they're each dancing to completely different songs.

- **Power Dynamics:** PMs are the big bosses who hold the purse strings and answer to the higher-ups. Supers? They're the boots on the ground, making sure things get done. But when the PM treats the Super like a mere minion, things get awkward faster than a middle school dance. Nobody likes being bossed around by someone who's never worn a hard hat.

- **Control Over Decisions:** Supers want to be able to make snap decisions—because, you know, things happen on-site in real-time. But if the PM wants to approve everything from what brand of nails to use, frustration is inevitable. At the same time, Supers can't exactly go rogue and start ordering helicopters without consulting the PM. It's all about trust, people!

4. Problem-Solving Styles

PMs and Supers solve problems like they're playing two entirely different video games—one's a strategy game, and the other's trying to beat the boss in real time.

- **Office-Based vs. Field-Based Perspectives:** PMs love data, reports, and 5-year plans. Supers? They just want to make sure the concrete sets before the rain hits. Sometimes the PM's "perfect on paper" solution doesn't fly when you're staring down a broken generator.

- **Proactive vs. Reactive Problem-Solving:** PMs pride themselves on being proactive, foreseeing problems like fortune-tellers in business suits. Supers, on the other hand, live for the adrenaline rush of putting out fires (not literally, we hope). The clash happens when the PM tries to prevent things that the Super is fully prepared to handle, leading to eye rolls from both sides.

5. Misalignment on Goals

While the PM is focused on the 10,000-foot view, the Super is knee-deep in mud trying to get the job done. It's like they're building the same house, but one's working from the penthouse, and the other's in the basement.

- **Focus on Project Scope vs. Day-to-Day Execution:** The PM's living in a world of client meetings, contract clauses, and PowerPoint slides. Meanwhile, the Super's just trying to get through today without a worker stepping on a nail. The challenge is getting these two realities to mesh without tearing apart the fabric of time and space.

- **Client vs. Site Concerns:** The PM says, "The client wants this done in three weeks!" The Super replies, "We'll be lucky if we're not still digging trenches in three weeks." When client demands meet the immovable force of real-world construction, sparks fly—and not the good kind.

Conclusion:

Let's face it, being a PM or a Super is tough enough on its own. Add in a pinch of misunderstanding, a dash of miscommunication, and a sprinkle of power struggle, and you've got a recipe for construction chaos. But with a little respect, some mutual understanding, and maybe a few fewer emails, PMs and Supers can work together like peanut butter and jelly—sticky, but ultimately, a winning combo.

Effective Communication Strategies

Effective communication is the secret sauce to any successful construction project—especially when keeping the peace between PMs and Supers. If these two don't communicate well, the project is headed straight for "delayed and over budget" territory. So, how can you build a harmonious PM-Super partnership? Let's dive into some communication strategies that will help you keep things smooth and—dare we say—pleasant.

1. Set Clear Communication Channels

If you think about it, PMs and Supers are like a construction site's odd couple. The PM is the planner, the Super is the doer, and without clear ways to communicate, things get messy. Imagine the chaos of trying to build a skyscraper with walkie-talkies that only work half the time. Here's how to avoid that:

- **Regular Meetings**: Sure, no one *loves* meetings, but in this case, they're a lifesaver. Have daily check-ins or weekly syncs where you discuss everything from progress to roadblocks. Think of these meetings as "relationship therapy" for your project.

- **Reports & Documentation**: You don't need a novel, but you do need key updates. Supers should send reports about the day's triumphs and disasters (and yes, they happen), while PMs should relay what's happening budget-wise and client-side. No one likes surprises unless they involve cake, so keep it concise.

- **Tech Tools**: Let's face it, no one has time to chase down email chains or sticky notes. Use a project management platform like Procore, where both of you can track updates, share documents, and, most importantly, avoid "he said, she said" moments.

Pro Tip: Establishing these communication channels early on is like laying the foundation for your project—do it right, and you'll avoid future headaches.

2. Balancing Transparency and Brevity

Let's talk about transparency. It's great... until it becomes *too* great. You don't need to know every detail about each other's day—just the important stuff. It's all about balance:

- **Transparency**: If you run into a problem, don't sweep it under the rug like leftover nails. Speak up. Whether it's a site issue or a budget hiccup, the sooner both parties know, the faster you can fix it.

- **Brevity**: That said, nobody has time for a five-page dissertation on why the site ran out of nails. Keep your updates focused on what *really* matters. Less is more. Remember, you're both busy people—not co-authors of a thriller novel.

Pro Tip: When in doubt, imagine explaining the update to a stranger in an elevator. If you can't do it in under a minute, it's probably too long.

3. Conflict Resolution Strategies

Like any dynamic duo, PMs and Supers will have disagreements. The question is: will you handle it like adults—or like two toddlers fighting over the last juice box?

- **Address Issues ASAP**: Don't let small issues snowball into giant boulders. The minute something feels off—whether it's scheduling, budget, or who stole the last donut—talk about it. Avoiding conflicts is like ignoring a fire alarm: sooner or later, something's going to burn.

- **Seek Common Ground**: At the end of the day, you both want the project to succeed. Try to remember that when you're in the heat of the moment. PMs may be fretting about the budget, while Supers care about quality and timelines. Work together to find a middle ground—it's like a construction tango.

- **Bring in the Referee**: If all else fails and you're at a standstill, it might be time to call in someone from senior management. Think of it as bringing in a referee to blow the whistle and making sure you both stay in bounds.

Pro Tip: Don't make it personal, it's not about who's *right*, it's about getting the project back on track.

4. Use Project Management Software Like a Boss

If PMs and Supers had a superhero sidekick, it'd be project management software. Platforms like Procore make it easy to track budgets, schedules, and progress—while keeping the lines of communication wide open. No more "I didn't get that email" excuses.

- **Procore**: This tool is like your project's Swiss Army knife. Both PMs and Supers can log in, update reports, and share notes—meaning everyone knows what's going on, all the time. Need a real-time update from the field? It's there. Wondering if the budget's holding up? It's a click away.

Pro Tip: Get the mobile app. You can check site updates while pretending to listen to that really long client meeting.

Conclusion

Effective communication between PMs and Supers isn't just a nice-to-have—it's essential. With clear channels, a balance between transparency and brevity, solid conflict resolution, and the right tech tools, you'll be well on your way to delivering projects on time, within budget, and with minimal hair-pulling.

Who knows? You might even become the dream team of construction.

Aligning Goals and Expectations: A PM and Super Love Story

Ah, the timeless dance between the PM and the Super. Two titans of construction, both with their own priorities, who somehow need to figure out how to share the same dance floor—without stepping on each other's toes. Misaligned goals and expectations can turn your beautiful construction project into a two-person wrestling match. But don't worry! With a few smooth moves, we can keep everyone waltzing to the same beat.

1. Defining Clear Project Goals from the Start: The Pre-Dance Pep Talk

Before any project kicks off, PMs and Supers must have "the talk." No, not *that* talk—the one where you sit down, lay out your hopes, dreams, and budget constraints, and figure out where you're both coming from. Because, let's be real—PMs are out there thinking, "We must not go over budget!" while Supers are all, "Quality is life!"

Like any good relationship, compromise is key. The PM is laser-focused on keeping the budget in check and hitting that sweet project deadline, while the Super is in the field managing people, machines, and probably swearing at supply chain delays. The goal? Agree on what really matters—whether it's spending a little extra time (and money) to avoid a patch job later or cutting corners to hit deadlines without making the building look like a Picasso painting.

Example: PM: "Look, we're two weeks behind, and the client's breathing down my neck. We need to push overtime." Super: "If you want it done right, I need time—not speed. Let's not pretend we can just microwave the foundation and call it good."

2. Establishing Key Performance Indicators (KPIs): Setting the Relationship GPS

Now that you've had "the talk," it's time to set some measurable goals to track your progress. These KPIs are like the construction project version of "Are we there yet?" Just like any good road trip, you need to know if you're on the right path or hopelessly lost somewhere between deadlines and budget reports.

Budget Tracking: Keep a sharp eye on how much you're spending, and remember—extra costs are like extra calories. They sneak up on you!

Schedule Adherence: You're going to be late sometimes, but tracking how late you are is key to making sure you're still getting home for dinner.

Quality Metrics: Count how many times you've had to redo something. If the number gets high enough, that's not a project, it's a do-over marathon.

Safety: "Safety first" isn't just a slogan. Keep tabs on those near-miss incidents because if your job site becomes a slapstick comedy show, everyone loses.

3. Regular Goal Check-ins: Couples Counseling, But for Construction

Even with all your KPIs and alignment chats, things can still get messy. That's why regular check-ins are important, kind of like construction couples' therapy. Every week or so, take the time to sit down and talk about how things are going.

Formal Check-ins: These are the times when you both put down your tools and say, "Alright, how's this relationship going?" The PM needs to hear what's happening on-site, and the Super needs to know if their labor force is about to get hit with budget cuts or client changes.

Identify Roadblocks: Maybe you're behind schedule because the weather decided to be as cooperative as a toddler at bedtime. Or maybe you've hit a snag with the supplier who thinks "just-in-time delivery" means "eventually." Either way, roadblocks are inevitable, and addressing them early keeps things moving forward.

Course Corrections: So, the project's a bit off-course. Do you panic? Nope. You adjust. Reallocate resources, swap out some tasks, maybe even bring in a third party to mediate (or at least to provide snacks).

4. Flexibility in Adjusting Goals: Go With the Flow—Construction Edition

Things change. Maybe it rains for a week, or maybe the client decides that the entire project needs to be redone because they suddenly hate bricks (clients are weird like that). Whatever the case, flexibility is your friend. Being rigid about goals and timelines is a great way to create tension and watch your project spiral into chaos. Instead, adjust your goals as you go and embrace the art of construction improv.

5. Proactive Communication: Talk Before Things Get Weird

Communication is like the secret sauce that keeps this whole PM-Super relationship from going off the rails. If the Super knows a storm is coming (literal or figurative), they should tell the PM. And if the PM knows the client is about to ask for 50,000 changes, the Super needs to hear it *before* it becomes a panic-inducing reality. Open, proactive communication is what prevents a construction project from turning into a game of telephone where everyone loses.

Conclusion: Dancing to the Same Beat

In the end, aligning goals and expectations is all about teamwork. PMs and Supers don't have to see eye to eye on everything—what's important is that you both know how to compromise, communicate, and adjust as needed. So, lace up your steel-toed boots, set those KPIs, and keep the dance going!

Building Trust and Mutual Respect: The PM-Super Power Dynamic

Trust and mutual respect between PMs and Supers are like the secret sauce that turns a construction project from a stress-fest into a well-oiled machine. Think of it like a buddy cop movie—both PMs and Supers have their strengths, but if they don't work together, the bad guys (delays, budget overruns, chaos) win. Let's explore how to build that perfect dynamic duo relationship between PMs and Supers.

1. The "You're Awesome, I'm Awesome" Approach

PMs and Supers are both essential, but they're playing different roles in the construction orchestra. Like a drummer and a guitarist, if they're not in sync, the whole band is off.

- **Recognizing Expertise:**
  PMs are the big-picture maestros—budgeting, scheduling, client schmoozing. Supers? They're the on-site rockstars—making sure the labor is in line, safety isn't just a suggestion, and walls are going up straight. When a PM acknowledges the Super's on-the-ground genius, it's like a "You're awesome" high-five. Supers, when you nod to the PM's skill

in juggling a million spreadsheets and a demanding client, it's your "I got you" moment.

- **Valuing Contributions:**
Let's face it, no one wants to feel like they're the sidekick. If a Super manages to get through a grueling week with no injuries and no delays, give them a shout-out. PMs, make sure to toss a "Nice save" when your Super fixes that late material delivery. Everyone loves a pat on the back, especially when it's for stopping the project from becoming a disaster movie.

2. Trust: It's Like Building a Bridge

Trust doesn't just appear; you have to build it, brick by brick. And yes, sometimes there's a lot of metaphorical dust, sweat, and swearing involved.

- **Open Dialogue:**
Ever had one of those days where you're both working late but avoiding each other like ex-roommates who didn't end on good terms? Don't. Instead, make open communication a habit. Weekly check-ins shouldn't feel like being sent to the principal's office. Think of it more like grabbing a coffee with a friend—you're just chatting about how to fix that "tiny" scheduling issue or that "slight" budget overrun.

- **Collaborative Problem-Solving:**
  When problems hit (because they always do), don't play the blame game. Instead, throw a "brainstorming party." If you're both stuck in a trench, wouldn't it be better to figure out how to climb out together than to argue over who pushed who? And hey, if one of you has the ladder, share!

3. Transparency: Let's Be Honest (Seriously)

Want to keep the respect rolling? Be transparent about the decisions that affect each other. You wouldn't suddenly change your road trip destination without telling your friend in the passenger seat, right?

- **Transparent Communication:**
  When PMs cut the budget or shift the schedule, Supers should know why. No one likes working in the dark—especially not when it's their job to ensure that the site is running smoothly. Be real about the reasons, and Supers, remember to ask questions. After all, a decision that looks good on paper can sometimes cause chaos in the real world (cue unexpected cement shortages).

- **Involving Supers in Planning:**
  Supers should get a seat at the table when the big decisions are made. Trust me, when the person on-site has no idea why certain deadlines were set, you're asking for trouble. Plus,

they might just have the one piece of advice that'll save you from a "what were we thinking" moment.

4. Public Praise: Who Doesn't Like a Gold Star?

Let's be real—everyone loves a bit of public recognition. When a Super nails the timeline or a PM navigates a budget crisis without making you cry, say it loud and say it proud.

- **Public Recognition:**
  Whether it's in a team meeting or in front of the client, give credit when credit's due. Bonus points for creative shout-outs—think "Superintendent of the Month" plaques or an honorary golden clipboard for the PM who kept everything under budget.

Conclusion

At the end of the day, building trust and respect between PMs and Supers is the key to making a project run smoothly—and maybe even making it enjoyable (gasp!). When you treat each other like the construction dream team you are, the whole site benefits. So, high-five your PM, nod to your Super, and remember that together, you can build skyscrapers—and avoid those budget-killing disaster movie moments.

Managing Budget, Time, and Quality Conflicts

How to Avoid a Construction Site Tug-of-War

In the construction world, you've got three main players in a high-stakes balancing act: Budget, Time, and Quality. And boy, do they love to fight. It's like trying to keep three toddlers happy with one cookie. On one side, you've got the PMs, fiercely guarding the budget and timelines like a hawk on a diet. On the other side, you have the Supers, keeping an eagle eye on quality and making sure no corner is cut (unless they say so). It's no surprise that sparks fly. So, how do we stop the mayhem and get everyone to play nice? Let's dive into some conflict resolution tips—construction style!

1. Techniques to Align the Budget-Obsessed PM with the Quality-Obsessed Super

(Otherwise known as "How to Get Along Without Throwing a Wrench at Each Other")

Integrated Planning Sessions:

Translation: Lock yourselves in a room, hash it out, and don't leave until you've agreed on something. Early on, sit down and plan the project together. PMs get to dream about budget savings, while

Supers map out how to make it all happen without duct tape. Remember, compromise isn't a dirty word!

Budget Flexibility:

Sure, it sounds like an oxymoron, but a little wiggle room in the budget can save everyone's sanity. Imagine the Super discovering some mystery pipes buried under the site that weren't on any plan. You'll be glad you had that rainy day fund. No one wants a project where the only thing on the up-and-up is the stress level.

Performance-Based Contracts:

Ah, nothing brings people together like money on the line. Tie performance to quality and timelines, and suddenly, everyone has skin in the game. The trick is making sure it's a fair race: clear quality standards, deadlines that don't feel like a sprint, and enough budget to make it happen without tears.

2. Case Studies in the Art of Dodging Disaster

(A.K.A. "Let's Learn from the Mistakes of Others")

Case Study 1: The Great Material Meltdown

A commercial building project hit a financial speed bump, and the PM thought cheaper materials were the answer. The Super, meanwhile, was having none of it. Fast forward to the end of this thrilling saga: they agreed to save costs where it wouldn't impact the building's stability, like opting for the economy version of flooring but keeping high-quality steel. Crisis averted, and nobody had to compromise their integrity (or the building's, for that matter).

Case Study 2: The Infrastructure Impasse

Community pressure had a PM chasing impossible deadlines like a greyhound after a rabbit. The Super, who values doing things right over doing them fast, saw disaster looming. They opted for a phased approach, taking care of the critical infrastructure repairs first and slowing down for the less urgent fixes. It's a classic case of *haste makes waste*—or worse, lawsuits.

3. Methods for Avoiding Budget-Quality Throwdowns

(Or "How to Keep the Peace on the Jobsite")

Open Communication Forums:

Think of this as couples therapy, but with more hard hats and fewer feelings. Set up regular chats to air grievances before they turn into full-blown battles. It's a safe space to say, "Hey, this budget cut is stressing me out," or "I know you want it done faster, but we need more hands to do it right."

Joint Decision-Making:

Two heads are better than one, right? Especially when they're not arguing. Get the PM and Super to make key decisions together. If you're cutting the budget, decide together which parts can afford a trim without the whole project falling apart like a house of cards.

Quality Control Checkpoints:

These are like pit stops in a race—time to pause, check how things are going, and make sure the wheels aren't about to fall off. Build-in checkpoints where both PMs and Supers review progress, ensuring everyone's on the same page. It's a lot easier to fix things before they've gone too far down the wrong road.

Creating a Culture of Accountability:

In other words, don't point fingers. Make sure both sides take responsibility for their part of the project. The PMs aren't the bad guys just because they're trying to keep things under budget, and Supers aren't villains for insisting on quality. Build a culture where everyone works toward the same goal: a project that's on time, on budget, and actually stands up when it's done.

Conclusion: The Ultimate Balancing Act

At the end of the day, managing budget, time, and quality is like juggling three chainsaws. Dangerous? Sure. Impossible? Not if you've got the right team and strategies in place. By planning together, communicating regularly, and remembering that everyone's in it for the same reason, PMs and Supers can make beautiful construction music together. Just remember: it's a marathon, not a sprint—and nobody wants to finish it bleeding money or fixing shoddy work.

Team Collaboration: PMs and Supers
The Dynamic Duo You Didn't Know You Needed

In the wild world of construction, it's like PMs and Supers are the Batman and Robin of every project. One handles the plans, budgets, and schedules; the other ensures things don't turn into a construction site version of *Jenga*. Getting these two to work together in perfect harmony is like orchestrating a Broadway show—challenging, yet magical when done right. Let's break down how to create this dream team leadership and why it's key to keeping your project from turning into a circus.

1. Strategies for PM and Super: From Frenemies to Besties

The construction crew can smell dysfunction from a mile away, so when the PM and Super are in sync, it creates a ripple effect that boosts morale and productivity. Here's how to make it happen:

- **Regular Joint Meetings (AKA "Construction Couple's Therapy"):** Schedule regular heart-to-hearts where both PMs and Supers can voice their opinions without throwing any wrenches (literally). This ensures everyone is on the same page before the chaos starts.

- **Shared Agenda:** Make sure you cover what really matters—budgets, safety, quality, and who's buying donuts for the next meeting. It's like a marriage, communication is key.

- **Coordinated Communication (One Voice, One Dream):** No one likes getting mixed messages, especially when it's about deadlines or safety issues. PMs and Supers should coordinate their updates like a well-rehearsed duet, so the team knows they're united.

  - **Joint Announcements:** Like parents telling their kids about bedtime (or budget cuts), make big announcements together to show the team you're both on the same side.

- **Mutual Support (BFFs, Even If It's Just for Show):** Back each other up in front of the team—even if you just had a heated debate about grout colors. The team needs to know that when it comes to decision-making, you've got each other's backs.

  - **Cross-Training:** Have PMs walk the site in their hard hats and Supers try their hand at spreadsheets. It builds empathy and helps you fake... I mean, *show* mutual respect in front of the team.

2. Managing Subcontractors, Vendors, and Clients—The Joint Approach

Handling subs, vendors, and clients is like juggling flaming chainsaws, but with the PM and Super teaming up, it's way less scary. Here's how to master the art of coordination:

- **Collaborative Planning (Two Heads Are Better Than One):** Sit down together to map out every subcontractor's schedule, vendor delivery, and client expectation. If one of you slips up, at least you'll both go down with the ship... I mean, you'll both be prepared to handle it.
    - **Involve Stakeholders:** Get subs and vendors in on the action early so they know the game plan. This also prevents anyone from crying foul later when the timeline shifts.
- **Setting Clear Expectations (Or As Clear As Construction Gets):** Lay down the law on what you expect from subs and vendors—both on quality and deadlines. PMs and Supers tag-teaming this effort means no one gets away with playing you against each other.
    - **Written Agreements:** Put everything in writing, because memories fade—especially when you're

dealing with 10 different moving parts and 100 emails a day.

- **Joint Performance Evaluations (And the Award for Best Sub Goes To...):** Sit down together to evaluate the subs and vendors. It's a great way to bond, and you'll make sure everyone's staying in line (without having to play good cop/bad cop).

3. Team Morale: When the Leadership Ship Doesn't Sink

When the PM and Super are in sync, the whole crew feels it. Think of it like a construction site dance—when leadership moves in rhythm, the team follows.

- **Boosting Morale (It's Not Just About Pizza Parties):** A unified leadership means a stable work environment where the crew doesn't feel like they're caught in a turf war. The team will follow your lead, so if you're on good terms, they'll likely be happier (and work harder).
    - **Positive Work Environment:** When the PM and Super aren't at each other's throats, the whole site is more relaxed, which can do wonders for productivity. You might even get through a whole day without someone losing their cool.

- **Effective Problem-Solving (The Power of Two):** When challenges pop up, and they will, a united PM and Super can tackle them with ease. The PM brings the budgets and spreadsheets, the Super brings the on-site know-how, and together they make sure things don't fall apart (literally).
    - **Consistent Messaging:** When things go wrong (and they will), the team needs to hear a unified message from leadership to keep everyone calm and focused.
- **Encouraging Accountability (Everyone's on the Hook):** When the leadership team is strong, the rest of the crew knows there's no room for slackers. The PM and Super can set the bar for accountability—and it doesn't hurt if they're seen high-fiving after a successful day.

Conclusion: PMs + Supers = Construction Magic

In the end, construction projects are a team effort, and when PMs and Supers are working together like a finely tuned machine, everything runs smoother. Whether it's managing subs, keeping morale high, or just making sure no one's stepping on each other's toes, it all comes down to unity. So, grab your partner, hit the site, and lead your team to construction victory—cape optional.

Technology in Project Management

Because Clipboards Are So Last Century

Let's face it: if you're still managing construction projects with a clipboard and a pen, you might as well throw in a pager and fax machine while you're at it. These days, technology is the name of the game. And when PMs and Supers join forces, armed with digital tools like Procore, they're basically the Avengers of the construction world. Minus the capes (though capes would be cool). Here's a humorous deep dive into how technology brings these superheroes of construction together, while navigating the ups, downs, and occasional "Help! I forgot my password!" moments.

1. How Modern Tools Improve Coordination Between PMs and Supers

Procore: The Swiss Army Knife for PMs and Supers

Think of Procore as the Swiss Army knife of project management software. It's got a tool for everything: communication, document sharing, task tracking, and even notifications (because who doesn't love getting *more* notifications, right?). When PMs and Supers use it, things get done faster—assuming, of course, everyone knows how to use it without getting lost in the software maze.

- **Centralized Communication**: Imagine a world where PMs and Supers don't have to send 10,000 texts, emails, and sticky notes to stay in sync. Procore creates a magical, centralized hub where everyone can share updates in real time, ensuring no one has to play "Telephone" with the team. Fewer mixed messages mean fewer "Wait, I thought you said Tuesday!" moments.

- **Task Management and Accountability**: No more awkward moments when PMs ask Supers, "Hey, did you finish that thing?" while Supers frantically dig through crumpled notes. With Procore, PMs assign tasks, and Supers check them off like a productivity game of bingo.

- **Document Management**: Picture this: PMs and Supers no longer lose sleep because a project plan was accidentally used as a coffee coaster. All documents live in the cloud, so everyone's on the same page. Literally.

- **Real-Time Updates**: Whether it's changes in project scope or realizing the wrong color paint is on the way, Procore sends real-time notifications. So Supers can say "Houston, we have a problem" *before* it becomes a real disaster.

- **Tool Integration**: Procore integrates with everything, except maybe your morning coffee. Accounting, scheduling, design—you name it, Procore's got a plugin for that. It's like getting a tech buffet, but without the food coma.

2. Challenges and Opportunities of Tech Adoption on Construction Sites

As much as technology is great, let's be real—there are a few bumps on the road to a digital construction utopia.

Challenges

- **Resistance to Change**: Some seasoned pros treat new technology like it's trying to take their lunch money. You'll hear, "I've been doing it this way for 30 years!" more than once. Convincing them to go digital can be like asking a cat to take a bath.

- **Learning Curve**: There's always that one guy whose first response to "Let's use the app" is, "What's an app?" Training is essential, or else you'll spend more time teaching than building.

- **Cost of Implementation**: Yes, good software isn't cheap, and when budgets are already tight, investing in tech can feel like asking the team to part with their favorite snacks. But hey, it's worth it in the long run—kind of like broccoli.

- **Data Security Concerns**: Construction's gone digital, but no one wants their site plans floating around the internet for all to see. Keeping that data locked down tighter than a vault at Fort Knox? That's a must.

Opportunities:

- **Increased Efficiency**: Technology automates all the boring stuff. You know, the paperwork, the endless task lists. That means PMs and Supers can focus on real issues, like where to get the best lunch in town.

- **Improved Collaboration**: When everyone's working off the same info, you spend less time chasing people for updates and more time making things happen. Plus, no more, "Did you get my email?" nonsense.

- **Data-Driven Insights**: Want to know if your project is headed for a budget overrun or if you're burning through materials too quickly? Technology can tell you—*before* you panic.

- **Enhanced Safety**: Construction sites are basically the Wild West without safety protocols. Tech helps prevent accidents by tracking everything from safety reports to training modules. Fewer injuries, more high-fives.

3. Leveraging Data for Real-Time Decision-Making

In construction, there's no such thing as a crystal ball (though it would be handy). Instead, we have data—lots of it. With the right tools, PMs and Supers can predict, adapt, and keep things running smoother than a freshly poured concrete slab.

- **Real-Time Analytics**: With real-time analytics, you can avoid the classic "Why didn't anyone tell me sooner?" meltdown. Dashboards track everything, so PMs and Supers can make decisions faster than you can say "change order."

- **Predictive Analytics**: If you're wondering if rain is going to ruin next week's concrete pour, predictive analytics has your back. Historical data helps PMs and Supers prepare for the unexpected—like a tech-savvy fortune teller.

- **KPIs**: Key performance indicators (KPIs) are the scorecards of construction. Want to know if you're ahead on budget or behind on schedule? The data's right there, and it's not afraid to tell the truth. No sugar-coating.

- **Streamlined Reporting**: Tired of spending hours putting together reports? Technology automates the whole process, so you can spend more time solving real problems—and less time wondering if you formatted that spreadsheet correctly.

- **Data Sharing**: When everyone has access to the same data, you eliminate those "he said, she said" arguments. Instead, it's just, "Look at the numbers, Bob. They don't lie."

Conclusion: Say Goodbye to the Stone Age

Let's face it—technology is here to stay, and it's making life easier for PMs and Supers (as long as they can remember their passwords). Sure, there are challenges, but the opportunities are too good to pass up. So whether you're boosting team collaboration, staying ahead of safety issues, or simply trying to figure out why the budget is giving you side-eye, embracing tech is the only way forward. Plus, it beats lugging around that clipboard.

Problem-Solving and Decision-Making

Construction's Comedy of Errors

In the world of construction, problem-solving and decision-making are like trying to fix a leaky pipe with duct tape—it can work, but only if you're using the right tools. PMs and Supers are like the Batman and Robin of construction, and their collaboration is key to ensuring everything runs smoothly. Spoiler alert: it rarely does without some hiccups. But don't worry! We've got structured approaches, prioritization tools, and collaborative techniques to save the day—without the need for superhero capes (unless you really want one).

1. Structured Approaches to Decision-Making: Because Wingin' It Never Works

Let's be honest, winging it might work when you're deciding on lunch, but when it comes to making decisions on a construction project? Not so much. A structured approach is like having a GPS for problem-solving—it tells you where to go without detours to "Oh no, I should have thought of that!"

a. Identify the Problem: AKA "Houston, We Have a Problem"

First, you need to know *what* you're solving. It's like trying to fix a squeaky door when the real issue is the roof leaking. Bring the PMs and Supers together and play a game of "Who's Got the Bigger Problem?"

b. Gather Relevant Information: AKA "What Are We Dealing With Here?"

Next, it's time to collect data. Think of this as detective work, minus the trench coat and magnifying glass. You're gathering project metrics, team input, and maybe even checking if that one guy is still using outdated blueprints.

c. Generate Alternatives: AKA "Let's Get Creative"

Here's where the magic happens: brainstorming. Let the ideas flow—no matter how crazy. Who knows, maybe using bubble wrap as insulation *could* work (just kidding… probably).

d. Evaluate Alternatives: AKA "Is This Really a Good Idea?"

Time to play construction jury. Evaluate each idea based on cost, time, and risk. Use fancy tools like decision matrices, or just do the ol' "pros vs. cons" list—whatever floats your construction boat.

e. Make the Decision: AKA "Can We All Agree on Something, Please?"

Once you've filtered out the bad ideas (sorry, bubble wrap), it's time for the team to come to a consensus. If no one agrees, a good ol' game of rock-paper-scissors may settle it (kidding again, mostly).

f. Implement the Decision: AKA "Now or Never"

You've made the call, so now it's time to make it happen! Assign roles, tell everyone what to do, and don't forget to bring donuts—it makes everything run smoother.

g. Review and Adjust: AKA "Wait, Did That Actually Work?"

After you've implemented the solution, check if it actually solved the problem. If not, don't panic—adjust as needed and maybe take a coffee break while you're at it.

2. Tools for Prioritizing Issues: Because You Can't Do Everything All at Once

Now that we've got structured decision-making down, it's time to figure out which issues to tackle first. You know, so you're not fixing the roof while the foundation is crumbling. Prioritization tools are here to make sure you're working smarter, not harder.

a. Eisenhower Matrix: AKA "Do I Have to Do This Right Now?"

This tool sorts tasks into four categories:

- **Urgent & Important**: Like that roof that's about to cave in, do it now.//
- **Important, Not Urgent**: Maybe the wallpaper in the lobby can wait a day.
- **Urgent, Not Important**: Delegate the coffee run.
- **Neither Urgent Nor Important**: Why are we even talking about this?

b. Impact vs. Effort Matrix: AKA "What's Worth the Sweat?"

With this tool, you figure out which tasks are worth the effort. If it's high impact and low effort—boom, do it now. High impact, high effort? Plan it out. And if it's low impact, high effort? Time to politely pass.

c. SWOT Analysis: AKA "What's the Worst That Could Happen?"

SWOT is all about figuring out the strengths, weaknesses, opportunities, and threats of a decision. Think of it as your construction crystal ball, minus the mystical part.

3. Collaborative Problem-Solving: Because Two (or More) Heads Are Better Than One

PMs and Supers are the dynamic duo, but working together is key. Collaboration means fewer mistakes and more high-fives when things go right.

a. Joint Problem-Solving Sessions: AKA "Let's All Just Get Along"

Set aside time to hash out issues together. It's like therapy for your project—everyone gets a turn to speak, no one gets interrupted, and hopefully, everyone leaves with a solution instead of a headache.

b. Brainstorming Techniques: AKA "Let the Creativity Flow"

Round-robin brainstorming or mind-mapping lets everyone contribute. No idea is too wild, and who knows, maybe the next big innovation in construction comes from these sessions. Just keep the ideas somewhat realistic.

c. Consensus Building: AKA "Can We All Agree on Something, Please?"

This involves some group voting or facilitated discussions. Think of it like picking the best pizza topping—everyone's got an opinion, but eventually, you'll reach a decision that works for (almost) everyone.

d. Role Reversal: AKA "Walk a Mile in My Steel-Toe Boots"

Have PMs and Supers step into each other's roles for a day. PMs will realize why Supers carry around giant rolls of plans, and Supers will understand why PMs drink so much coffee.

e. Continuous Feedback Loop: AKA "Let's Keep Talking"

Communication is key. Set up regular feedback sessions where PMs and Supers can vent, suggest improvements, and keep things moving forward without too much drama.

Conclusion: Teamwork Makes the Construction Dream Work

Effective problem-solving and decision-making aren't just essential; they're the secret sauce to keeping your project on track. When PMs and Supers join forces using structured approaches, prioritization tools, and collaboration techniques, you get more than just a finished project—you get a masterpiece. Now, go grab that Swiss Army knife (aka Procore) and get to work!

Conflict Resolution and Mediation

Because You Can't Just Send PMs and Supers to Timeout

Conflict is as unavoidable on a construction site as coffee spills on paperwork. When PMs and Supers butt heads, it can lead to tension worse than that morning traffic jam. But don't worry, this section is here to help you sort through the chaos of conflict like a well-organized toolbox, with just the right strategies to avoid project disasters. And yes, we've got case studies, because what's conflict without some juicy real-world drama?

1. How to Mediate Conflicts Between PMs and Supers (Before the Project Explodes)

Mediating PM vs. Super conflicts is like being the referee in a game where both teams forgot the rules. The goal? Keep the peace before someone storms off-site with a dramatic "I'm done!" Here's how to channel your inner peacemaker:

a. Identify the Conflict Early

You know that awkward tension in the air when the PM and Super won't look each other in the eye during meetings? Yeah, that's your cue. Spotting these conflicts early is key—before they turn into full-blown site meltdowns. Regular check-ins and airing grievances over coffee (or something stronger) might just save your project.

b. Foster Open Communication

Think of this like marriage counseling, but with hard hats. Both sides need to vent, so make sure:

- You listen like a therapist (nod, smile, and don't pick sides).
- Everyone feels heard, even when what they're saying sounds like "blah, blah, blah."
- The conversation stays on track—this is about the job, not who's hogging the best parking spot.

c. Clarify Misunderstandings

You'd be amazed how many conflicts come down to "I thought *you* were doing that!" versus "No, I thought *you* were!" Sit them down, grab a whiteboard, and map out who's responsible for what—like chore charts, but for adults.

d. Focus on Interests, Not Positions

Don't let them lock horns over trivial stuff like that tiny budget line item. Instead, remind them that they're on the same team: The *Let's Finish This Project On Time and Under Budget* team.

e. Explore Solutions Together

Now that we've all calmed down (fingers crossed), it's brainstorming time. PMs and Supers are both pretty clever, so let them come up with the solutions—just make sure their ideas don't include things like "Let's move this to next year."

f. Document Agreements

No one remembers what they agreed to a month ago, so write it down. Treat it like the Holy Grail of peace agreements, something to wave around if tensions flare again.

g. Follow Up

Just because they've stopped glaring at each other doesn't mean you're in the clear. Regular check-ins are like relationship maintenance, make sure both parties are still playing nice.

2. Third-Party Mediation: When Things Get Too Spicy

Sometimes the PM and Super just can't settle things on their own, and you're stuck playing the middleman while they battle it out. Enter the neutral third party: the construction world's version of a relationship counselor.

a. Neutral Third-Party Mediation

When the tension's too thick to cut with a saw, bring in someone impartial. They'll set ground rules (like "No eye-rolling, Jerry") and guide the conversation so that nobody ends up using a blueprint as a weapon.

b. Structured Mediation Sessions

These sessions are like therapy, but with hard hats. Private one-on-ones, joint venting, and finally, some handshakes (or at least the coldest of nods). The mediator's job is to find common ground—and that doesn't mean where to build the new breakroom.

c. When to Involve Upper Management

If things have escalated to "We're never working together again," it's time to pull in the big guns—upper management. But they're not here to bark orders (hopefully); they're here to show that teamwork makes the dream work, even if the dream involves countless RFIs and punch lists.

Conclusion: Conflict Resolution—The Secret Sauce to Project Success

Resolving PM vs. Super conflicts is about as crucial to a project as, well, everything else. Communication, collaboration, and a good mediator (with some snacks on hand) can go a long way. Plus, as long as you keep an eye on things and learn from past hiccups, you'll be able to dodge those big disputes and keep the project (and your sanity) intact.

Leadership Development for PMs and Superintendents

Becoming the Captain of the Construction Ship (Without Sinking It!)

Effective leadership is like the secret sauce in a burger—without it, things fall apart. Whether you're a PM steering the budget ship or a Super making sure that ship doesn't capsize on-site, both roles need some fine-tuned leadership skills. In this section, we'll explore how to develop the skills that'll keep your construction ship sailing smoothly. You'll also learn about the importance of mentorship, cross-training, and the never-ending joy of continuous improvement. Ready? Grab your hard hat, because we're about to dive in!

1. Leadership Skills: What Both PMs and Supers Need to Learn (Preferably Yesterday)

While PMs and Supers may have different roles, they both need some of the same leadership skills to avoid chaos on-site and, more importantly, to avoid having team members "accidentally" throw their phones at them.

a. Communication Skills

- *Clarity and Conciseness*: Look, nobody has time for an epic speech. Both PMs and Supers need to communicate as clearly and concisely as possible. Pretend you're giving instructions to a toddler—but one who can handle a jackhammer.

- *Active Listening*: This is the "nod and look concerned" skill. Seriously though, it's about actually paying attention to what your team says, not just nodding like you're trying to win an Oscar.

b. Decision-Making and Problem-Solving

- *Analytical Thinking*: You're the Sherlock Holmes of construction, minus the hat and British accent. Analyze the situation, weigh your options, and make decisions that don't involve building something upside down.

- *Collaborative Problem-Solving*: Let's be honest, no one wants to be the only one holding the "what-do-we-do-now" bag. Involve the team in problem-solving. Two heads are better than one (unless both heads are arguing over where the porta-potties should go).

c. Emotional Intelligence (EQ)

- *Self-Awareness*: Know thyself. If you're frustrated, take a deep breath before yelling at the forklift.

- *Empathy*: You're not just managing tasks, you're managing humans (and their feelings—yes, even in construction!). A little understanding goes a long way.

d. Conflict Resolution

- *Mediation Skills*: Be the construction Gandhi. When tempers flare, cool it down before someone "accidentally" drives a skid steer over someone's lunch.

- *Negotiation Skills*: Negotiating is like playing chess—except the chess pieces are humans, and you're trying to get them to agree on who's paying for that extra 300 yards of concrete.

e. Adaptability and Resilience

- *Flexibility*: When you're working on a project, expect the unexpected. And when the unexpected happens, don't just throw your hard hat in frustration—adjust the plan.

- *Resilience*: Keep calm and construction on. A positive attitude can keep the whole team from feeling like they're sinking in a quagmire of delays and rework.

f. Team Building and Motivation

- *Inspiring Trust and Respect*: Lead by example. Show your team you've got their back (and occasionally buy them donuts).

- *Recognition and Rewards*: Let's face it, people love kudos. So give it out generously—especially when they avoid hitting the power line with the backhoe.

2. Mentorship and Cross-Training: Because Learning the Hard Way is Overrated

a. Mentorship Programs

- *Pairing Veterans with Newbies*: No, not hazing mentoring. Experienced PMs and Supers should take rookies under their wing and teach them the ways of the construction force.

- *Relationship Building*: Mentorship isn't just about work—it's about building relationships. Make it a two-way street. Maybe that rookie knows the one trick to stop your smartphone from crashing every time you open Procore.

b. Cross-Training Initiatives

- *Role Exchange*: Have your PMs and Supers swap places for a day. It's like wife-swap but way more useful. PMs get to see the chaos on-site, and Supers realize how fun it is to sit in meetings all day.

- *Holistic Perspective*: The more everyone understands each other's roles, the less likely anyone is to say, "That's not my job" when the next mini-disaster strikes.

3. Continuous Improvement and Learning: Because You're Never Done Learning (Even If You Wish You Were)

a. Ongoing Training Programs

- *Workshops and Seminars*: Take your leaders to workshops and seminars, not just for the snacks, but to actually learn new leadership tricks.

- *Certifications*: Encourage them to get certifications like the PMP or OSHA training. If nothing else, it gives them more acronyms to throw around during meetings.

b. Encouraging Feedback and Reflection

- *Performance Reviews*: No one loves them, but they're useful—like kale. Make reviews about growth and improvement, not just about whether that last concrete pour was on time.

- *Culture of Continuous Improvement*: The goal is always getting better—kinda like leveling up in a video game, but without the joystick.

c. Utilizing Technology for Learning

- *Online Learning Platforms*: Sure, YouTube is great for watching cat videos, but it's also handy for leadership training. Provide access to platforms that keep leaders sharp (and maybe still allow a cat video or two for stress relief).

- *Knowledge Sharing Tools*: Use platforms like Slack or Teams for more than just memes. Share articles, tips, and tricks. (But seriously, still share the memes too.)

Conclusion: The Construction Leadership Revolution (Or At Least a Minor Overhaul)

Leadership development is the key to turning a good project into a great one. By giving PMs and Supers the right tools, mentorship, and learning opportunities, you're not just building better leaders—you're building better projects. Plus, happy leaders make for happy teams, which means fewer hard hats are thrown in anger. And isn't that the dream?

Building Long-Term Relationships

How to Keep the Peace Between PMs and Supers Without Needing a Mediator

Let's be honest: fostering a long-term relationship between Project Managers PMs and Supers is a bit like maintaining a good marriage—it takes work, patience, and a fair amount of negotiation over who gets to control the remote. But when done right, these professional partnerships don't just survive—they thrive, leading to smoother projects, happier teams, and fewer workplace "divorces."

1. How PMs and Supers Can Foster Long-Term Professional Relationships (Without Needing Couples Therapy)

a. **Open Communication and Transparency (AKA: "Just Tell Me What You Really Mean")**

- *Regular Check-Ins*: Think of these like date nights, but for work. Set up regular check-ins where you can both talk about your feelings... about the project. Whether it's over a formal meeting or a quick coffee run, keeping the lines of communication open ensures that any grievances are handled *before* you start passive-aggressively emailing each other.

- *Transparency in Decision-Making*: PMs, if you're moving the budget around like a game of Monopoly, just let the Supers know *why*. Supers, if you're changing the schedule again, explain the method behind the madness. Transparency builds trust. Or at least prevents side-eye in meetings.

b. **Collaboration on Goals and Expectations (AKA: "We're In This Together")**

- *Aligning Project Objectives*: Nothing says teamwork like actually being on the same page. Make sure you agree on project goals from the get-go, otherwise, it's like trying to build a house with two different blueprints. Spoiler: it doesn't end well.

- *Shared Accountability*: Celebrate wins together and share the responsibility when things go sideways. When the project succeeds, throw a pizza party. When it fails, maybe... well, also throw a pizza party? At least you'll be disappointed *together*.

c. **Building Personal Connections (AKA: "Let's Be Work Friends")**

- *Investing Time in Relationship Building*: Spend time getting to know each other beyond project deadlines and budgets. Who knows, you might find out that your PM also loves golf or that your Super is really into… golf. Or, you know, whatever people do outside of work.

- *Respecting Each Other's Expertise*: PMs know their Gantt charts and Supers know their way around the job site. Recognize that both of you bring something important to the table—and that table is probably covered in blueprints and coffee cups.

2. The Benefits of Strong, Enduring Collaborations (AKA: "Why Being BFFs With Your Super Isn't a Bad Idea")

a. **Improved Communication and Coordination (AKA: "We Can Finish Each Other's... Sentences")**

- *Streamlined Processes*: Long-term relationships mean less paperwork and more of that sweet, sweet shorthand that only comes with knowing someone's quirks. You'll get to the point where you can resolve problems with a single nod, a raised eyebrow, or an emoji.

- *Shared Understanding*: Once you've worked with the same PM or Super long enough, you'll know exactly what they need before they ask. It's basically telepathy. But, like, with job orders.

b. **Increased Efficiency and Productivity (AKA: "Less Time Arguing, More Time Building")**

- *Reduced Learning Curve*: When you've been through the trenches together, you already know each other's strengths and weaknesses. It's like that scene in every buddy cop movie where they finally start to trust each other and everything clicks. Just without the car chases.

- *Proactive Problem Solving*: When you've got a solid working relationship, problems don't fester. You tackle them head-on like a dynamic duo, cutting down issues before they become full-blown disasters.

c. **Enhanced Project Outcomes (AKA: "Look What We Did!")**

- *Consistency in Quality*: When PMs and Supers work well together, the project runs smoothly and consistently. It's like a well-oiled machine, except with less grease and more spreadsheets.

- *Higher Client Satisfaction*: When you're not constantly bickering, the clients notice. They appreciate a well-run project—and happy clients mean repeat business. And repeat business means you get to work together again! Lucky you.

3. Strategies for Retaining Talented PMs and Supers (AKA: "Keeping the Dream Team Together")

a. Providing Opportunities for Growth (AKA: "Keep Learning or Get Bored")

- *Professional Development*: The secret to keeping talented PMs and Supers? Keep them engaged. Give them workshops, certifications, and maybe even send them to that leadership retreat in the mountains where they'll learn about trust falls and synergy.

- *Career Pathing*: No one wants to feel like they're stuck in a dead-end job. Help your PMs and Supers see the bright future ahead. Career growth is like fertilizer for talent—wait, that sounded weird. You get the idea.

b. Recognizing and Rewarding Contributions (AKA: "High-Fives All Around")

- *Celebrating Successes*: Did you just finish a project under budget and on time? Break out the champagne—or, more realistically, the donuts. A little recognition goes a long way.

- *Feedback and Appreciation*: Don't wait until the yearly review to say "thanks" or "great job." Give each other props regularly. You'll be amazed how far a little appreciation can go in making people feel valued.

c. **Fostering a Positive Work Environment (AKA: "Because Work Shouldn't Suck")**

- *Encouraging Work-Life Balance*: Look, no one likes a workaholic. Encourage each other to take time off and unplug. You'll come back to the job site refreshed and less likely to lose it over the next change order.

- *Creating a Supportive Culture*: Build a culture where PMs and Supers help each other out. It's like being on a team, except without the weird matching uniforms.

Conclusion (AKA: "We're Better Together")

Building long-term relationships between PMs and Supers isn't just a nice-to-have—it's essential for getting things done without constant headaches. Open communication, collaboration, and a genuine personal connection can turn a potentially stressful partnership into a productive (and maybe even fun) one. So, invest in the relationship, celebrate the wins, and keep working together. Before you know it, you'll be the dynamic duo everyone else in the industry is jealous of.

Towards Better Project Outcomes

The PM-Super Peace Treaty

Well, folks, we've made it to the end of this "survival guide" for PMs and Supers—or, if you prefer, our group therapy session for construction teams. It's been a journey, but the message is clear: the PM-Super relationship is the secret sauce to any successful project. Whether you're battling conflicting priorities or trying to get on the same page with goals, the key to keeping the peace is good ol' fashioned communication, trust, and a healthy dose of mutual respect. So, before you run back to your spreadsheets and job sites, let's recap the nuggets of wisdom we've uncovered and gaze into a future where PMs and Supers don't just tolerate each other—they thrive together.

Recap of Key Takeaways for Building Stronger PM-Super Relationships (AKA: "Stuff You Should Probably Remember")

- **Clear Communication is Non-Negotiable:** Whether it's a formal meeting, a quick chat in the hallway, or a desperate text at 2 a.m. about why the concrete wasn't poured, keeping the lines of communication open is non-negotiable. It's not just about talking; it's about making sure you're both in sync on what really matters, like what time lunch is, I mean, project priorities, expectations, and daily tasks.

- **Respect Each Other's Expertise:** PMs, before you roll your eyes at the Super's constant scheduling changes, remember they're juggling on-the-ground chaos like labor issues, safety concerns, and quality control. Supers, before you grumble about the PM's love of spreadsheets, remember: those numbers and budgets keep the lights on. Bottom line? Respect each other's roles. It's like a buddy-cop movie, everyone's got their own strengths.

- **Collaborate, Don't Compete:** Sure, it's tempting to keep score—who's hitting the budget, who's keeping the timeline, and who's winning the argument over where the porta-potties go. But here's a pro tip: you're both on the same team. When you collaborate instead of competing, everyone wins. And if all else fails, remember when the project succeeds, you both get to take credit (and maybe a celebratory drink).

Encouragement to Invest in Relationships for the Benefit of the Project and Company (AKA: "The Secret to Not Hating Each Other")

Let's face it—building a strong relationship between PMs and Supers isn't just about making your workday tolerable. It's about setting up the entire project for success. When you're on the same wavelength, you avoid costly mistakes, reduce rework, and—dare we say it—hit those impossible deadlines. Not only that, but your teamwork can boost morale across the whole crew. And let's be real, when subcontractors and laborers see that you're not about to rip each other's heads off, they tend to be a lot more motivated to get the job done.

For the company? Well, strong PM-Super relationships translate to happy clients, successful projects, and—cha-ching!—better profits. So, it's not just about surviving this project, it's about setting the stage for a future full of wins.

Vision for the Future of PM-Super Collaborations in the Construction Industry (AKA: "We're All in This Together")

Looking ahead, the construction industry is speeding into a future full of fancy technology, tighter deadlines, and clients who demand everything yesterday. PMs and Supers who figure out how to work together will not only survive these changes—they'll thrive. We're entering an era where collaboration, both between humans and with technology, is the name of the game.

Tools like Procore are making it easier than ever for PMs and Supers to communicate, share data, and avoid the dreaded "I didn't know about that" situation. And with a stronger focus on leadership development, there's no reason why the PM-Super relationship can't become a model of teamwork, respect, and high-fives all around.

At the end of the day, the projects that truly succeed are built with more than just steel and concrete—they're built on the trust, collaboration, and mutual respect of the teams making it happen. So, let's face the future, tackle the challenges, and move toward a world where PMs and Supers are more like partners than adversaries. After all, it's a lot more fun when we're all in this together.

Now, go forth and build—just maybe check in with each other before the next change order.

www.ingramcontent.com/pod-product-compliance
Lightning Source LLC
Chambersburg PA
CBHW050250220526
45465CB00002B/626